Water
76

感谢蜱虫
Ticks for Life

Gunter Pauli

［比］冈特·鲍利 著

［哥伦］凯瑟琳娜·巴赫 绘

隋淑光 译

上海远东出版社

丛书编委会

主　任：田成川

副主任：何家振　闫世东　林　玉

委　员：李原原　翟致信　靳增江　史国鹏　梁雅丽

　　　　任泽林　陈　卫　薛　梅　王　岢　郑循如

　　　　彭　勇　王梦雨

特别感谢以下热心人士对童书工作的支持：

匡志强　宋小华　解　东　厉　云　李　婧　庞英元

李　阳　刘　丹　冯家宝　熊彩虹　罗淑怡　旷　婉

杨　荣　刘学振　何圣霖　廖清州　谭燕宁　王　征

李　杰　韦小宏　欧　亮　陈强林　陈　果　寿颖慧

罗　佳　傅　俊　白永喆　戴　虹

目录

Contents

在非洲中部热
带草原上的一个小池
塘里，一条鱼正在游泳。
一只孤独的蝌蚪也在游来
游去，他快要变成青
蛙了。

A fish is
swimming in a small
pool of water in the
middle of the African
savanna. A lonely tadpole
swims around. He is about
to turn into a frog.

一只蝌蚪在游来游去……

A tadpole swims around ...

我是怎么来到这里的?

How did I ever get here?

"你注意到附近
没有河吗？" 蝌蚪问道。
"我为什么要关心那个？我在这里
过得很快乐。" 鱼回答道。
"嗯，我很好奇我是怎么来到这里的，
还有，你是怎么来到这里的？这附
近没有我们的族群。"

"Did you know
that there is no river
anywhere close by?" asks the
tadpole.
"Why should I care? I am quite
happy here," responds the fish.
"Well, I wonder, how did I ever
get here? And, how did you get
here? There is no family
around."

"肯定是我的
爸爸妈妈送我来的，可能
我是个淘气的孩子。"
"你会飞吗？"
"我是鱼，又不是鸟。你
能飞吗？"

"I am sure
my parents sent me
off. Maybe I was a very
naughty kid."
"Can you fly?"
"I am a fish, not a bird.
Can you fly?"

我是鱼，又不是鸟

I am a fish, not a bird

......被鹳鸟吃掉了！

... eaten by a stork!

"不，根本不
行，我只是一只蝌蚪。我在
找我的爸爸妈妈和兄弟姐妹，但是
没有他们的任何痕迹，甚至连骨架都
没有。"
"他们很可能被鹳鸟吃掉了，我
知道鹳鸟喜欢吃青蛙。"

"No, not at all, I
am just a tadpole. But I
have been looking around for
my mom and dad, brothers and
sisters. There is no sign of them,
nothing, not even a skeleton."
"They were most likely eaten
by a stork! I know they love
to eat frogs."

"他们也喜欢吃
鱼。那么，如果我们的爸爸
妈妈不在附近，是谁带我们飞到
这儿的？"

"哦，你认为自己是坐喷气式飞
机来的？真是异想天开。"

"And they love
to eat fish as well.
So if our parents are not
around, who flew us here?"
"Oh, now you think that
you arrived on a jet
plane? You are such a
dreamer."

是谁带我们飞到这儿的？

Who flew us here?

在他飞往欧洲的途中……

On his way to Europe …

"也许不是飞
机，可能是一只鹳鸟在某处觅
食，那时你我也许还是一只卵，被
嵌进了他的脚趾缝里。在他飞往欧洲的
途中，他在这个池塘里驻足休息，我们
就被留下来了。"

"很有趣的想法，但是这个池
塘是谁挖的呢？"

"Maybe not a
plane, but perhaps a
stork was having a meal
somewhere. It could be that you
and I, as eggs, were stuck to his
feet. On his way to Europe, he
made a pit stop here at this pool
and we got off."

"Interesting thought. But
who dug this pool?"

"你看到过大
象、犀牛和水牛洗尘土
澡吗？"
"噗，当他们试图通过洗尘土澡来
刮掉那些蜱虫的时候，尘土飞
扬，你什么都看不到。"

"Have you ever seen
an elephant, a rhino or a
buffalo take a dust bath?"
"Pffff, you cannot see anything
when he tries to scratch his
butt and back clean to get
rid of those ticks."

选尘土澡

A dust bath

微小的、黑乎乎的生物

Tiny, black creatures

"你是说那些微小
的、黑乎乎的、样子丑陋、有坚
硬头部的吸血生物吗？"

"我们很幸运，我们生活在水里，而那些害
虫不能在这里生存。"

"那为什么大象和水牛不能通过洗澡来摆脱
这些讨厌的小动物呢？"

"因为一年中的大部分时间里，这
个池塘里没有水。"

"You mean those
tiny, black, ugly-looking,
hard-headed, bloodsucking
creatures?"

"We are so lucky. We live in water
where these pests don't survive."

"So why don't the elephants and
buffaloes take a nice soaking bath,
and drown the little menaces?"

"Because most of the year
there is no water in
this pool."

"但是那意味着池
塘会干涸！我们也会死的！"

"那就是生命循环。大象洗尘土澡时，
用身体压出一个布满灰尘的干坑，然后雨水
会填满它，再后来鹳鸟会带来新的生命。那就
是这个故事告诉我们的。"

"也就是说，此刻我们之所以能享受生命
的快乐，应该感谢蜱虫？"

……这仅仅是开始……

"But that means this
pool will dry out! And we will
then die!"

"That is the cycle of life. The
elephants take a dry bath, create a
solid dust pool, the rain will fill it, and
the stork will bring new life. That is what
this fable tells us."

"And this means we have the ticks
to thank for this moment of joy?"

... AND IT HAS ONLY
JUST BEGUN!...

……这仅仅是开始！……

… AND IT HAS ONLY JUST BEGUN! …

蜱虫依靠吸血来生存，如果找不到寄主就会死亡。

Ticks need blood to survive. Ticks unable to find a host will die.

蜱虫有四对足。幼虫阶段只有三对，第四对要在第一次吸血后才开始发育。

Ticks have four pairs of legs. Larval ticks have only three pairs of legs; the fourth pair only grows after it has had its first meal of blood.

The first pair of legs can smell, sense vibration, notice increased moisture and feel the body heat of potential hosts.

第一对足的功能有嗅闻、感觉振动、分辨湿度变化，以及感受潜在寄主的体温。

Ticks love a warm and humid climate, as they need moisture in the air to undergo a metamorphosis from a larva to a nymph and then to a full-grown tick. Low temperatures inhibit their growth.

蜱虫喜欢温暖潮湿的气候，因为它在从幼虫变为若虫，再变为成虫时，需要吸收空气中的水分。低温会抑制它的发育。

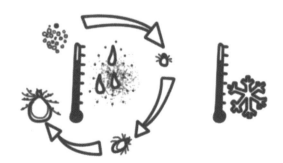

除了剧毒的DDT以外，尚不知道有其他化学药品能杀死蜱虫。珍珠鸡是最高效的蜱虫"终结者"，几只珍珠鸡就可以把一公顷土地上的蜱虫消灭干净。

No known chemicals except the very toxic DDT kill ticks. Guinea fowl are the most successful tick eliminators. Just a few guinea fowl will clear a hectare of all ticks.

蜱虫寄生在大象身上，安居在皮肤的褶皱里，或者通过伤口进入血液。大象洗尘土澡是为了摆脱蜱虫。

Ticks infest elephants and nestle into the wrinkles of the skin or enter through a cut to get to the blood. Elephants shower with dust and sand to dislodge ticks.

The little oxpecker cleans the rhino by picking off ticks, but the bird prefers thick ticks full of blood, ignoring the little ones that are just as much of a nuisance.

牛椋鸟能帮助犀牛清理蜱虫，但是它偏爱那些吸饱了血的胖胖的蜱虫，会忽略那些同样有威胁的瘦家伙。

The stork flies from the Western Cape (South Africa) to Europe in 26 days and returns in 49 days. Storks use the lift of air thermals to gain height and glide to their destination.

鹳鸟从南非西开普省飞到欧洲要花26天时间，返程时要花49天。它利用空气热气流升空，升到一定高度后，滑翔到目的地。

Would you ever consider a tick as a friend?

你会把蜱虫看作朋友吗?

How do you think the fish and the frog ended up in this isolated pool of water?

你怎么看待鱼和青蛙最终来到这片偏僻的池塘这件事?

水牛、犀牛和大象的身体重到洗个尘土澡就能挖出一个池塘吗?

Are buffaloes, rhinos and elephants heavy enough to dig a pool by taking a dust bath?

Do you like the idea that life emerges again every year again from a dust pool that fills with water during the rainy season, but turns into a dust pool again in the summer?

在雨季蓄满雨水、夏季干涸的池塘里,生命年复一年地出现,你喜欢这样吗?

It is springtime. Go and look for a pool of water and search for tadpoles. Take a net with you and catch a few. Keep them in clean water. Feed them lettuce and leftovers. If the tadpole is still very young, cook the lettuce first, as that makes it easier to eat. It can take 6 to 12 weeks before a tadpole turns into a frog and you will be able to observe all the stages in the development of the tadpole. If you want you can catch a small fish and watch it grow too. Find one that is not a carnivore, so that the tadpole and the fish can happily live alongside each other. Once they grow big, release them back into nature and enjoy the memories!

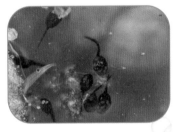

春天来了。到池塘里找一找蝌蚪，用网捞几只，把它们养在净水里，用生菜和剩饭剩菜喂它们。如果蝌蚪还很小，就把生菜煮熟再喂，以便它们进食。蝌蚪要花6—12周时间才能变成青蛙，你可以观察到整个发育过程。如果你想捉一条小鱼来观察它的发育，试着捉一条非肉食性的鱼，这样它就能与蝌蚪和平共处了。一旦它们长大了，将它们放回自然界（原先抓它们的地方），享受喂养它们的美好回忆吧。

学科知识
Academic Knowledge

生物学	我们所认为的有害物种在生态系统中也有作用；寄生虫和病菌承担"生态系统工程师"的角色；寄生植物和寄生动物在养分循环中的作用及其对生物多样性的影响；寄生生物促进物种间遗传物质的转移，并引起了生物进化；鸟类必须快速地定居、摄食和消化，在长期的进化中它们失去了牙齿，颚部也退化了；在准备迁徙时，为方便飞行，鹳鸟会把生殖器官的重量降低到准备生殖时的1%；水是影响生物生存的重要因素。
化 学	蟑虫通过产生抗凝血肽来防止血液堵塞，这对医药工业很有启发。
物 理	鹳鸟通过拍动翅膀升空并飞到目的地，它们所消耗的能量是那些利用热气流升空并飞行的鸟类的23倍多；经过干旱季节后土壤变得致密，雨水很难渗入。
工程学	利用水的压力来清洗物体表面的方式被称作喷流，利用砂的压力清洗物体表面的方式被称作喷砂。
经济学	商业具有季节性，在一年时间内消费行为会有变化。
伦理学	根据好和坏来快速区分物种的做法，并没有真正理解每一个现存物种所承担的角色和责任，我们需要花时间来理解每一个现存物种在每一种生态背景下的价值。
历 史	鹳鸟从开始迁徙到返回的时间和人类从女性排卵到婴儿诞生的时间（9个月）一致，因此在维多利亚时期有"一只鹳鸟带来一个婴儿"的故事，而在古希腊神话中则有鹳鸟偷婴儿的传说。
地 理	鹳鸟在迁徙时要避开大面积的水域，因此从南非到欧洲的路线要么是飞越黎巴嫩（东线），要么是飞越摩洛哥（西线），因为大面积的水域上空的空气中没有可供它们自由上升的热气流。
数 学	鹳鸟飞往欧洲和返回时所花的时间不同，前者需要26天，后者需要49天，这是由风、气流和食物三个因素决定的。
生活方式	在伊索寓言中首次提到了"狐狸和鹳鸟"的故事，然后安徒生进行了再创作；父母常用这些故事来向小孩解释婴儿是如何来到世上的，而不必深入讲生育的细节。
社会学	在一些亚洲文化中，时间被认为是往复的，失去的还可以回来；而西方文化则认为时间是线性的，流逝的就永远流逝了。
心理学	每个人都问过"我是从哪里来的"这个问题，并向父母、长辈和同龄人寻求答案，从这一问题出发可以进一步质疑生命以及我们的父母是从哪里来的；寻根使我们明确生活的目的并且有稳定感。
系统论	生命网中有很多纵横交错的联系，它们看起来似乎是随机的、偶然的，但是一旦深入研究就会对这种联系和连接产生更好的理解，这对维持自然界中生存和死亡的循环是非常有必要的。

情感智慧
Emotional Intelligence

鱼

鱼非常不愿意和蝌蚪进行深入交流，对于蝌蚪提出的更深层次的问题，她快速简单地作答，语气冷淡甚至有些讽刺的意味，似乎是想终止这个话题。但是蝌蚪接连不断的提问最终吸引了她的注意力，她开始和蝌蚪讨论起她们生活环境的起源问题，即这个偏僻的小池塘是她们仅有的栖息地。蝌蚪的话触动了她的心弦，使她开始认真对待这场谈话。于是鱼开始分享自己对生命周期的见解，她对自身起源和未来的深入理解使蝌蚪感到惊讶。在这个故事将要结束时，鱼的认识越来越接近问题的核心，并开始进行哲学思考。

蝌　蚪

蝌蚪好奇自己是怎么来到这里的，并开始寻求答案。但她除了和共处于这块栖息地的鱼探讨外别无选择，因此她没有考虑鱼的感受，就觉得有必要讨论这个问题。鱼有些粗暴的回答没有使她泄气，她坚持追问她们是从哪里来的。蝌蚪认可鱼对她的生存困境的判断，她准备深入思考一些问题，这促使她和鱼进行了第一次接触。她认识到一个可怕的事实，即她鄙视的一些物种却和她的生存密不可分。然而，现实情况使她焦虑，她意识到应该对蜉虫表示感谢，接受尽管她希望能活着，但她的生命最终会结束这一现实。

艺术
The Arts

你看到过特写镜头下的蜉虫吗？找一些蜉虫头部的照片，你会觉得似乎看到了外星人。在A4纸上画一画这张照片，用不同的颜色突出头部或者四对足上最不可思议的部位。你现在看到的是一个让人产生不适并且能传播疾病的生物，但它同时又是生物多样性最有效的传播者之一。

思维拓展
Systems: Making the Connections

生物界充满了惊喜。从蜱虫的故事能联想到未知的生命起源和进化。我们都知道生态系统复杂多样，但是仍然对于微小的寄生生物竟和地球上的大型哺乳动物联系在一起而感到震惊。这个故事能引导我们理解候鸟的特殊作用。蜱虫和非洲的五种大型动物（水牛、犀牛、大象、狮子和豹）之间的关联，一条鱼、一只青蛙、一只鹳鸟和一只蜱虫的关联，是对生命之网的真实揭示。我们想要弄清生命是如何起源和进化的，尽管我们知道自己将永远无法完全掌握现实。这激励我们从哲学层面思考我们是如何出现在地球上的，以及我们为什么会生存在现有的条件下。这个故事告诉我们，生命是有限的，而生命的循环和延续是无限的。微小的寄生生物对其他生物的影响强化了这样的观念——我们必须尊重所有形式的生物，即使是那些微小的、我们还没有理解的生物。生物多样性越丰富，我们的生活就会越美好，就会充溢着我们还没意识到的机会。

动手能力
Capacity to Implement

你注意到了身边的生命之网了吗？站在至少10个人组成的圈里，用手捏紧一个线团的线头，每当某个人能说出一个和前面的人所说的词语关联的词，就把线团传递给他。可以从你想到的任何一个词语（也许是与你家有关的词）开始这个游戏。每个人要逼自己至少想出50种关联，想的时候要打开思路，不要局限于纯科学，可以扩展到神话、文化象征，甚至纯想象，只要你能说服其他人认同这种想象即可。

故事灵感来自
This Fable Is Inspired by

彼得·雷蒙多
Peter Raimondo

自从 2003 年起，彼得·雷蒙多一直作为指导老师在位于南非赫卢赫卢韦国家禁猎区的野外领导能力学校工作。他曾经在开普敦大学就读，主要学习哲学、人类学和环境地理科学，但是他更喜欢在户外工作，向人们提供获得直观知识和认识自然规律的机会。他组织了穿越南非荒野的徒步活动，让参加者在没有舒适的帐篷、预加工食品以及通信设施的条件下体验生活。

图书在版编目（CIP）数据

冈特生态童书.第三辑修订版:全36册:汉英对照 /
(比)冈特·鲍利著;(哥伦)凯瑟琳娜·巴赫绘;
何家振等译.—上海:上海远东出版社,2022
书名原文:Gunter's Fables
ISBN 978-7-5476-1850-9

Ⅰ.①冈… Ⅱ.①冈… ②凯… ③何… Ⅲ.①生态环
境–环境保护–儿童读物—汉、英 Ⅳ.①X171.1-49

中国版本图书馆CIP数据核字(2022)第163904号
著作权合同登记号图字09-2022-0637号

策　　划 张　蓉
责任编辑 祁东城
封面设计 魏　来李　廉

冈特生态童书

感谢蜱虫

[比]冈特·鲍利　著
[哥伦]凯瑟琳娜·巴赫　绘

隋淑光　译

记得要和身边的小朋友分享环保知识哦！
八喜冰淇淋祝你成为环保小使者！